COMPANY NAME

Name	
Address	
Email	
Phone Number	
Fax Number	

LOG BOOK DETAILS

Log Start Date	
Log book Number	

Index

Year:

January	February	March
1	1	1
2	2	2
3	3	3
4	4	4
5	5	5
6	6	6
7	7	7
8	8	8
9	9	9
10	10	10
11	11	11
12	12	12
13	13	13
14	14	14
15	15	15
16	16	16
17	17	17
18	18	18
19	19	19
20	20	20
21	21	21
22	22	22
23	23	23
24	24	24
25	25	25
26	26	26
27	27	27
28	28	28
29		29
30		30
31		31

Year:

April	May	June
1	1	1
2	2	2
3	3	3
4	4	4
5	5	5
6	6	6
7	7	7
8	8	8
9	9	9
10	10	10
11	11	11
12	12	12
13	13	13
14	14	14
15	15	15
16	16	16
17	17	17
18	18	18
19	19	19
20	20	20
21	21	21
22	22	22
23	23	23
24	24	24
25	25	25
26	26	26
27	27	27
28	28	28
29	29	29
30	30	30
	31	

Year:

July	August	September
1	1	1
2	2	2
3	3	3
4	4	4
5	5	5
6	6	6
7	7	7
8	8	8
9	9	9
10	10	10
11	11	11
12	12	12
13	13	13
14	14	14
15	15	15
16	16	16
17	17	17
18	18	18
19	19	19
20	20	20
21	21	21
22	22	22
23	23	23
24	24	24
25	25	25
26	26	26
27	27	27
28	28	28
29	29	29
30	30	30
31	31	

Year:

October	November	December
1	1	1
2	2	2
3	3	3
4	4	4
5	5	5
6	6	6
7	7	7
8	8	8
9	9	9
10	10	10
11	11	11
12	12	12
13	13	13
14	14	14
15	15	15
16	16	16
17	17	17
18	18	18
19	19	19
20	20	20
21	21	21
22	22	22
23	23	23
24	24	24
25	25	25
26	26	26
27	27	27
28	28	28
29	29	29
30	30	30
31		

Year:

January	February	March
1	1	1
2	2	2
3	3	3
4	4	4
5	5	5
6	6	6
7	7	7
8	8	8
9	9	9
10	10	10
11	11	11
12	12	12
13	13	13
14	14	14
15	15	15
16	16	16
17	17	17
18	18	18
19	19	19
20	20	20
21	21	21
22	22	22
23	23	23
24	24	24
25	25	25
26	26	26
27	27	27
28	28	28
29		29
30		30
31		31

Year:

April	May	June
1	1	1
2	2	2
3	3	3
4	4	4
5	5	5
6	6	6
7	7	7
8	8	8
9	9	9
10	10	10
11	11	11
12	12	12
13	13	13
14	14	14
15	15	15
16	16	16
17	17	17
18	18	18
19	19	19
20	20	20
21	21	21
22	22	22
23	23	23
24	24	24
25	25	25
26	26	26
27	27	27
28	28	28
29	29	29
30	30	30
	31	

Year:

July	August	September
1	1	1
2	2	2
3	3	3
4	4	4
5	5	5
6	6	6
7	7	7
8	8	8
9	9	9
10	10	10
11	11	11
12	12	12
13	13	13
14	14	14
15	15	15
16	16	16
17	17	17
18	18	18
19	19	19
20	20	20
21	21	21
22	22	22
23	23	23
24	24	24
25	25	25
26	26	26
27	27	27
28	28	28
29	29	29
30	30	30
31	31	

Year:

October	November	December
1	1	1
2	2	2
3	3	3
4	4	4
5	5	5
6	6	6
7	7	7
8	8	8
9	9	9
10	10	10
11	11	11
12	12	12
13	13	13
14	14	14
15	15	15
16	16	16
17	17	17
18	18	18
19	19	19
20	20	20
21	21	21
22	22	22
23	23	23
24	24	24
25	25	25
26	26	26
27	27	27
28	28	28
29	29	29
30	30	30
31		

SHEEP RECORD

Date Added:	Sheep Id:
Name:	D. O. B.:

Sex:	Birth Type:		Breed:
Type:	◯ Lamb ◯ Ewe ◯ Ram		Genotype:

Acquisition Details

Purchased From:	Date
Purchased address:	

Purpose of Acquisition: ◯Breeding   ◯Raising   ◯Sale   ◯Other:

Date Applied for Identifiers:	Purchase Price:

Weight Details	At Birth	Day 30	Day 60	6 Months	9 Month	1 Year

Pedigree Details

Sheep's Picture

Notes_____

_____

SHEEP RECORD

Date Added:		Sheep Id:	
Name:		D. O. B.:	
Sex:	Birth Type:		Breed:
Type:	○ Lamb ○ Ewe ○ Ram		Genotype:

Acquisition Details

Purchased From:		Date
Purchased address:		

Purpose of Acquisition: ○ Breeding   ○ Raising   ○ Sale   ○ Other:

Date Applied for Identifiers:		Purchase Price:

Weight Details	At Birth	Day 30	Day 60	6 Months	9 Month	1 Year

Pedigree Details

Sheep's Picture

Notes_____

_____

SHEEP RECORD

Date Added:	Sheep Id:
Name:	D. O. B.:
Sex:	Birth Type:	Breed:
Type:	○ Lamb ○ Ewe ○ Ram	Genotype:

Acquisition Details

Purchased From:	Date
Purchased address:	

Purpose of Acquisition: ○Breeding ○Raising ○Sale ○Other:

Date Applied for Identifiers: | Purchase Price:

Weight Details	At Birth	Day 30	Day 60	6 Months	9 Month	1 Year

Pedigree Details

Sheep's Picture

Notes _____

_____

SHEEP RECORD

Date Added:		Sheep Id:
Name:		D. O. B.:
Sex:	Birth Type:	Breed:
Type:	○ Lamb ○ Ewe ○ Ram	Genotype:

Acquisition Details

Purchased From:		Date
Purchased address:		

Purpose of Acquisition: ○ Breeding   ○ Raising   ○ Sale   ○ Other:

Date Applied for Identifiers: Purchase Price:

Weight Details	At Birth	Day 30	Day 60	6 Months	9 Month	1 Year

Pedigree Details

Sheep's Picture

Notes_____

SHEEP RECORD

Date Added:	Sheep Id:	
Name:	D. O. B.:	
Sex:	Birth Type:	Breed:
Type: ○ Lamb ○ Ewe ○ Ram		Genotype:

Acquisition Details

Purchased From:	Date
Purchased address:	

Purpose of Acquisition: ○Breeding   ○Raising   ○Sale   ○Other:

Date Applied for Identifiers:	Purchase Price:

Weight Details	At Birth	Day 30	Day 60	6 Months	9 Month	1 Year

Pedigree Details

Notes_____

_____

SHEEP RECORD

Date Added:		Sheep Id:	
Name:		D. O. B.:	
Sex:	Birth Type:		Breed:
Type:	○ Lamb ○ Ewe ○ Ram		Genotype:

Acquisition Details

Purchased From:		Date
Purchased address:		

Purpose of Acquisition: ○ Breeding   ○ Raising   ○ Sale   ○ Other:

Date Applied for Identifiers:		Purchase Price:

Weight Details	At Birth	Day 30	Day 60	6 Months	9 Month	1 Year

Pedigree Details

Sheep's Picture

Notes_____

SHEEP RECORD

Date Added:	Sheep Id:
Name:	D. O. B.:
Sex:	Birth Type:	Breed:
Type: ○ Lamb ○ Ewe ○ Ram	Genotype:

Acquisition Details

Purchased From:	Date
Purchased address:	

Purpose of Acquisition: ○Breeding   ○Raising   ○Sale   ○Other:

Date Applied for Identifiers: Purchase Price:

Weight Details	At Birth	Day 30	Day 60	6 Months	9 Month	1 Year

Pedigree Details

Sheep's Picture

Notes_____

_____

SHEEP RECORD

Date Added:		Sheep Id:	
Name:		D. O. B.:	
Sex:	Birth Type:		Breed:
Type:	◯ Lamb ◯ Ewe ◯ Ram		Genotype:

Acquisition Details

Purchased From:		Date
Purchased address:		

Purpose of Acquisition: ◯Breeding   ◯Raising   ◯Sale   ◯Other:

Date Applied for Identifiers: Purchase Price:

Weight Details	At Birth	Day 30	Day 60	6 Months	9 Month	1 Year

Pedigree Details

Sheep's Picture

Notes_____

_____

SHEEP RECORD

Date Added:		Sheep Id:	
Name:		D. O. B.:	
Sex:	Birth Type:		Breed:
Type:	◯ Lamb ◯ Ewe ◯ Ram		Genotype:

Acquisition Details

Purchased From:		Date
Purchased address:		

Purpose of Acquisition: ◯Breeding   ◯Raising   ◯Sale   ◯Other:

Date Applied for Identifiers:	Purchase Price:

Weight Details	At Birth	Day 30	Day 60	6 Months	9 Month	1 Year

Pedigree Details

Sheep's Picture

Notes_____

_____

SHEEP RECORD

Date Added:		Sheep Id:	
Name:		D. O. B.:	
Sex:	Birth Type:		Breed:
Type:	○ Lamb ○ Ewe ○ Ram		Genotype:

Acquisition Details

Purchased From:		Date
Purchased address:		

Purpose of Acquisition: ○Breeding   ○Raising   ○Sale   ○Other:

Date Applied for Identifiers:	Purchase Price:

Weight Details	At Birth	Day 30	Day 60	6 Months	9 Month	1 Year

Pedigree Details

Sheep's Picture

Notes

SHEEP RECORD

Date Added:		Sheep Id:	
Name:		D. O. B.:	
Sex:	Birth Type:		Breed:
Type:	○ Lamb ○ Ewe ○ Ram		Genotype:

Acquisition Details

Purchased From:	Date
Purchased address:	

Purpose of Acquisition: ○ Breeding   ○ Raising   ○ Sale   ○ Other:

Date Applied for Identifiers:	Purchase Price:

Weight Details	At Birth	Day 30	Day 60	6 Months	9 Month	1 Year

Pedigree Details

Sheep's Picture

Notes_____

_____

SHEEP RECORD

Date Added:		Sheep Id:	
Name:		D. O. B.:	
Sex:	Birth Type:		Breed:
Type:	○ Lamb ○ Ewe ○ Ram		Genotype:

Acquisition Details

Purchased From:		Date
Purchased address:		

Purpose of Acquisition: ○ Breeding   ○ Raising   ○ Sale   ○ Other:

Date Applied for Identifiers: | Purchase Price:

Weight Details	At Birth	Day 30	Day 60	6 Months	9 Month	1 Year

Pedigree Details

Sheep's Picture

Notes_____

SHEEP RECORD

Date Added:	Sheep Id:
Name:	D. O. B.:

Sex:	Birth Type:	Breed:
Type:	○ Lamb ○ Ewe ○ Ram	Genotype:

Acquisition Details

Purchased From:	Date
Purchased address:	

Purpose of Acquisition: ○Breeding   ○Raising   ○Sale   ○Other:

Date Applied for Identifiers:	Purchase Price:

Weight Details	At Birth	Day 30	Day 60	6 Months	9 Month	1 Year

Pedigree Details

Sheep's Picture

Notes_____

_____

SHEEP RECORD

Date Added:		Sheep Id:
Name:		D. O. B.:
Sex:	Birth Type:	Breed:
Type:	○ Lamb ○ Ewe ○ Ram	Genotype:

Acquisition Details

Purchased From:	Date
Purchased address:	

Purpose of Acquisition: ○Breeding   ○Raising   ○Sale   ○Other:

Date Applied for Identifiers:	Purchase Price:

Weight Details	At Birth	Day 30	Day 60	6 Months	9 Month	1 Year

Pedigree Details

Sheep's Picture

Notes_____

_____

SHEEP RECORD

Date Added:	Sheep Id:	
Name:	D. O. B.:	
Sex:	Birth Type:	Breed:
Type: ◯ Lamb ◯ Ewe ◯ Ram		Genotype:

Acquisition Details

Purchased From:		Date
Purchased address:		

Purpose of Acquisition: ◯Breeding ◯Raising ◯Sale ◯Other:

Date Applied for Identifiers: _____ Purchase Price:

Weight Details	At Birth	Day 30	Day 60	6 Months	9 Month	1 Year

Pedigree Details

Sheep's Picture

Notes_____

_____

SHEEP RECORD

Date Added:		Sheep Id:	
Name:		D. O. B.:	
Sex:	Birth Type:		Breed:
Type:	⚪ Lamb ⚪ Ewe ⚪ Ram		Genotype:

Acquisition Details

Purchased From:		Date
Purchased address:		

Purpose of Acquisition: ⚪Breeding ⚪Raising ⚪Sale ⚪Other:

Date Applied for Identifiers:	Purchase Price:

Weight Details	At Birth	Day 30	Day 60	6 Months	9 Month	1 Year

Pedigree Details

Sheep's Picture

Notes_____

_____

SHEEP RECORD

Date Added:		Sheep Id:	
Name:		D. O. B.:	
Sex:	Birth Type:	Breed:	
Type:	○ Lamb ○ Ewe ○ Ram	Genotype:	

Acquisition Details

Purchased From:		Date	
Purchased address:			

Purpose of Acquisition: ○ Breeding   ○ Raising   ○ Sale   ○ Other:

Date Applied for Identifiers:	Purchase Price:

Weight Details	At Birth	Day 30	Day 60	6 Months	9 Month	1 Year

Pedigree Details

Sheep's Picture

Notes_____

_____

SHEEP RECORD

Date Added:		Sheep Id:	
Name:		D. O. B.:	
Sex:	Birth Type:		Breed:
Type:	○ Lamb ○ Ewe ○ Ram		Genotype:

Acquisition Details

Purchased From:		Date
Purchased address:		

Purpose of Acquisition: ○ Breeding   ○ Raising   ○ Sale   ○ Other:

Date Applied for Identifiers:	Purchase Price:

Weight Details	At Birth	Day 30	Day 60	6 Months	9 Month	1 Year

Pedigree Details

Sheep's Picture

Notes_____

_____

29

SHEEP RECORD

Date Added:		Sheep Id:	
Name:		D. O. B.:	
Sex:	Birth Type:		Breed:
Type:	○ Lamb ○ Ewe ○ Ram		Genotype:

Acquisition Details

Purchased From:		Date
Purchased address:		

Purpose of Acquisition: ○Breeding   ○Raising   ○Sale   ○Other:

Date Applied for Identifiers:		Purchase Price:

Weight Details	At Birth	Day 30	Day 60	6 Months	9 Month	1 Year

Pedigree Details

Sheep's Picture

Notes_____

_____

SHEEP RECORD

Date Added:		Sheep Id:
Name:		D. O. B.:

Sex:	Birth Type:	Breed:

Type: ○ Lamb ○ Ewe ○ Ram | Genotype:

Acquisition Details

Purchased From:	Date

Purchased address:

Purpose of Acquisition: ○ Breeding ○ Raising ○ Sale ○ Other:

Date Applied for Identifiers: | Purchase Price:

Weight Details	At Birth	Day 30	Day 60	6 Months	9 Month	1 Year

Pedigree Details

Sheep's Picture

Notes_____

SHEEP RECORD

Date Added:	Sheep Id:
Name:	D. O. B.:
Sex:	Birth Type:	Breed:
Type:	○ Lamb ○ Ewe ○ Ram	Genotype:

Acquisition Details

Purchased From:	Date
Purchased address:	

Purpose of Acquisition: ○Breeding ○Raising ○Sale ○Other:

Date Applied for Identifiers: Purchase Price:

Weight Details	At Birth	Day 30	Day 60	6 Months	9 Month	1 Year

Pedigree Details

Sheep's Picture

Notes _____

_____

SHEEP RECORD

Date Added:	Sheep Id:
Name:	D. O. B.:

Sex:	Birth Type:	Breed:
Type: ◯ Lamb ◯ Ewe ◯ Ram		Genotype:

Acquisition Details

Purchased From:	Date
Purchased address:	

Purpose of Acquisition: ◯Breeding ◯Raising ◯Sale ◯Other:

Date Applied for Identifiers: _____ Purchase Price:

Weight Details	At Birth	Day 30	Day 60	6 Months	9 Month	1 Year

Pedigree Details

Sheep's Picture

Notes_____

_____

SHEEP RECORD

Date Added:		Sheep Id:	
Name:		D. O. B.:	
Sex:	Birth Type:		Breed:
Type:	○ Lamb ○ Ewe ○ Ram		Genotype:

Acquisition Details

Purchased From:		Date
Purchased address:		

Purpose of Acquisition: ○Breeding   ○Raising   ○Sale   ○Other:

Date Applied for Identifiers:		Purchase Price:

Weight Details	At Birth	Day 30	Day 60	6 Months	9 Month	1 Year

Pedigree Details

Sheep's Picture

Notes_____

_____

SHEEP RECORD

Date Added:		Sheep Id:
Name:		D. O. B.:
Sex:	Birth Type:	Breed:
Type:	◯ Lamb ◯ Ewe ◯ Ram	Genotype:

Acquisition Details

Purchased From:		Date
Purchased address:		

Purpose of Acquisition: ◯Breeding   ◯Raising   ◯Sale   ◯Other:

Date Applied for Identifiers: | Purchase Price:

Weight Details	At Birth	Day 30	Day 60	6 Months	9 Month	1 Year

Pedigree Details

Sheep's Picture

Notes_____

_____

SHEEP RECORD

Date Added:	Sheep Id:
Name:	D. O. B.:

Sex:	Birth Type:	Breed:
Type:	○ Lamb ○ Ewe ○ Ram	Genotype:

Acquisition Details

Purchased From:	Date
Purchased address:	

Purpose of Acquisition: ○ Breeding   ○ Raising   ○ Sale   ○ Other:

Date Applied for Identifiers:	Purchase Price:

Weight Details	At Birth	Day 30	Day 60	6 Months	9 Month	1 Year

Pedigree Details

Sheep's Picture

Notes_____

_____

SHEEP RECORD

Date Added:		Sheep Id:	
Name:		D. O. B.:	
Sex:	Birth Type:		Breed:
Type:	○ Lamb ○ Ewe ○ Ram		Genotype:

Acquisition Details

Purchased From:		Date
Purchased address:		

Purpose of Acquisition: ○Breeding ○Raising ○Sale ○Other:

Date Applied for Identifiers:		Purchase Price:

Weight Details	At Birth	Day 30	Day 60	6 Months	9 Month	1 Year

Pedigree Details

Sheep's Picture

Notes_____

_____

SHEEP RECORD

Date Added:		Sheep Id:
Name:		D. O. B.:
Sex:	Birth Type:	Breed:
Type:	○ Lamb ○ Ewe ○ Ram	Genotype:

Acquisition Details

Purchased From:	Date
Purchased address:	

Purpose of Acquisition: ○Breeding   ○Raising   ○Sale   ○Other:

Date Applied for Identifiers:	Purchase Price:

Weight Details	At Birth	Day 30	Day 60	6 Months	9 Month	1 Year

Pedigree Details

Sheep's Picture

Notes_____

_____

SHEEP RECORD

Date Added:	Sheep Id:
Name:	D. O. B.:

Sex:	Birth Type:	Breed:
Type: ○ Lamb ○ Ewe ○ Ram		Genotype:

Acquisition Details

Purchased From:	Date

Purchased address:

Purpose of Acquisition: ○Breeding ○Raising ○Sale ○Other:

Date Applied for Identifiers:	Purchase Price:

Weight Details	At Birth	Day 30	Day 60	6 Months	9 Month	1 Year

Pedigree Details

Sheep's Picture

Notes_____

_____

SHEEP RECORD

Date Added:		Sheep Id:
Name:		D. O. B.:
Sex:	Birth Type:	Breed:
Type:	◯ Lamb ◯ Ewe ◯ Ram	Genotype:

Acquisition Details

Purchased From:	Date
Purchased address:	

Purpose of Acquisition: ◯Breeding   ◯Raising   ◯Sale   ◯Other:

Date Applied for Identifiers:	Purchase Price:

Weight Details	At Birth	Day 30	Day 60	6 Months	9 Month	1 Year

Pedigree Details

Sheep's Picture

Notes_____

_____

SHEEP RECORD

Date Added:		Sheep Id:	
Name:		D. O. B.:	
Sex:	Birth Type:		Breed:
Type:	○ Lamb ○ Ewe ○ Ram		Genotype:

Acquisition Details

Purchased From:		Date
Purchased address:		

Purpose of Acquisition: ○Breeding   ○Raising   ○Sale   ○Other:

Date Applied for Identifiers:		Purchase Price:

Weight Details	At Birth	Day 30	Day 60	6 Months	9 Month	1 Year

Pedigree Details

Sheep's Picture

Notes_____

_____

SHEEP RECORD

Date Added:	Sheep Id:	
Name:	D. O. B.:	
Sex:	Birth Type:	Breed:
Type: ○ Lamb ○ Ewe ○ Ram		Genotype:

Acquisition Details

Purchased From:	Date
Purchased address:	

Purpose of Acquisition: ○Breeding   ○Raising   ○Sale   ○Other:

Date Applied for Identifiers:	Purchase Price:

Weight Details	At Birth	Day 30	Day 60	6 Months	9 Month	1 Year

Pedigree Details

Sheep's Picture

Notes_____

_____

SHEEP RECORD

| Date Added: | Sheep Id: |
| Name: | D. O. B.: |

| Sex: | Birth Type: | Breed: |
| Type: | ◯ Lamb   ◯ Ewe   ◯ Ram | Genotype: |

Acquisition Details

| Purchased From: | Date |
| Purchased address: | |

Purpose of Acquisition: ◯Breeding   ◯Raising   ◯Sale   ◯Other:

| Date Applied for Identifiers: | Purchase Price: |

Weight Details	At Birth	Day 30	Day 60	6 Months	9 Month	1 Year

Pedigree Details

Sheep's Picture

Notes_____

_____

SHEEP RECORD

Date Added:	Sheep Id:
Name:	D. O. B.:

Sex:	Birth Type:	Breed:

Type:	○ Lamb ○ Ewe ○ Ram	Genotype:

Acquisition Details

Purchased From:	Date

Purchased address:

Purpose of Acquisition: ○Breeding   ○Raising   ○Sale   ○Other:

Date Applied for Identifiers:	Purchase Price:

Weight Details	At Birth	Day 30	Day 60	6 Months	9 Month	1 Year

Pedigree Details

Sheep's Picture

Notes_____

_____

SHEEP RECORD

Date Added:		Sheep Id:	
Name:		D. O. B.:	
Sex:	Birth Type:		Breed:
Type:	◯ Lamb ◯ Ewe ◯ Ram		Genotype:

Acquisition Details

Purchased From:		Date
Purchased address:		

Purpose of Acquisition: ◯Breeding   ◯Raising   ◯Sale   ◯Other:

Date Applied for Identifiers:	Purchase Price:

Weight Details	At Birth	Day 30	Day 60	6 Months	9 Month	1 Year

Pedigree Details

Sheep's Picture

Notes_____

_____

SHEEP RECORD

Date Added:		Sheep Id:	
Name:		D. O. B.:	
Sex:	Birth Type:	Breed:	
Type:	○ Lamb ○ Ewe ○ Ram	Genotype:	

Acquisition Details

Purchased From:		Date
Purchased address:		

Purpose of Acquisition: ○Breeding ○Raising ○Sale ○Other:

Date Applied for Identifiers:	Purchase Price:

Weight Details	At Birth	Day 30	Day 60	6 Months	9 Month	1 Year

Pedigree Details

Sheep's Picture

Notes_____

_____

SHEEP RECORD

Date Added:	Sheep Id:
Name:	D. O. B.:

Sex:	Birth Type:	Breed:
Type:	○ Lamb ○ Ewe ○ Ram	Genotype:

Acquisition Details

Purchased From:	Date
Purchased address:	

Purpose of Acquisition: ○Breeding   ○Raising   ○Sale   ○Other:

Date Applied for Identifiers: | Purchase Price:

Weight Details	At Birth	Day 30	Day 60	6 Months	9 Month	1 Year

Pedigree Details

Sheep's Picture

Notes_____

_____

47

SHEEP RECORD

Date Added:	Sheep Id:	
Name:	D. O. B.:	
Sex:	Birth Type:	Breed:
Type: ○ Lamb ○ Ewe ○ Ram		Genotype:

Acquisition Details

Purchased From:	Date
Purchased address:	

Purpose of Acquisition: ○Breeding ○Raising ○Sale ○Other:

Date Applied for Identifiers: _____ Purchase Price:

Weight Details	At Birth	Day 30	Day 60	6 Months	9 Month	1 Year

Pedigree Details

Notes_____

_____

SHEEP RECORD

Date Added:	Sheep Id:
Name:	D. O. B.:

Sex:	Birth Type:	Breed:
Type: ◯ Lamb ◯ Ewe ◯ Ram		Genotype:

Acquisition Details

Purchased From:	Date
Purchased address:	

Purpose of Acquisition: ◯Breeding ◯Raising ◯Sale ◯Other:

Date Applied for Identifiers:	Purchase Price:

Weight Details	At Birth	Day 30	Day 60	6 Months	9 Month	1 Year

Pedigree Details

Sheep's Picture

Notes_____

_____

SHEEP RECORD

Date Added:		Sheep Id:	
Name:		D. O. B.:	
Sex:	Birth Type:	Breed:	
Type:	◯ Lamb ◯ Ewe ◯ Ram	Genotype:	

Acquisition Details

| Purchased From: | Date |
| Purchased address: | |

Purpose of Acquisition: ◯Breeding   ◯Raising   ◯Sale   ◯Other:

| Date Applied for Identifiers: | Purchase Price: |

Weight Details	At Birth	Day 30	Day 60	6 Months	9 Month	1 Year

Pedigree Details

Sheep's Picture

Notes_____

_____

SHEEP RECORD

Date Added:			Sheep Id:	
Name:			D. O. B.:	
Sex:		Birth Type:		Breed:
Type:	○ Lamb ○ Ewe ○ Ram			Genotype:

Acquisition Details

Purchased From:		Date
Purchased address:		

Purpose of Acquisition: ○Breeding ○Raising ○Sale ○Other:

Date Applied for Identifiers:		Purchase Price:

Weight Details	At Birth	Day 30	Day 60	6 Months	9 Month	1 Year

Pedigree Details

Sheep's Picture

Notes_____

_____

SHEEP RECORD

Date Added:		Sheep Id:	
Name:		D. O. B.:	
Sex:	Birth Type:	Breed:	
Type:	○ Lamb ○ Ewe ○ Ram	Genotype:	

Acquisition Details

| Purchased From: | | Date |
| Purchased address: | | |

Purpose of Acquisition: ○Breeding   ○Raising   ○Sale   ○Other:

| Date Applied for Identifiers: | Purchase Price: |

Weight Details	At Birth	Day 30	Day 60	6 Months	9 Month	1 Year

Pedigree Details

Sheep's Picture

Notes_____

_____

SHEEP RECORD

Date Added:		Sheep Id:	
Name:		D. O. B.:	
Sex:	Birth Type:		Breed:
Type:	○ Lamb ○ Ewe ○ Ram		Genotype:

Acquisition Details

Purchased From:		Date
Purchased address:		

Purpose of Acquisition: ○Breeding ○Raising ○Sale ○Other:

Date Applied for Identifiers:		Purchase Price:

Weight Details	At Birth	Day 30	Day 60	6 Months	9 Month	1 Year

Pedigree Details

Sheep's Picture

Notes_____

SHEEP RECORD

Date Added:		Sheep Id:
Name:		D. O. B.:
Sex:	Birth Type:	Breed:
Type:	◯ Lamb ◯ Ewe ◯ Ram	Genotype:

Acquisition Details

Purchased From:	Date
Purchased address:	

Purpose of Acquisition: ◯ Breeding   ◯ Raising   ◯ Sale   ◯ Other:

Date Applied for Identifiers:	Purchase Price:

Weight Details	At Birth	Day 30	Day 60	6 Months	9 Month	1 Year

Pedigree Details

Sheep's Picture

Notes_____

_____

SHEEP RECORD

Date Added:		Sheep Id:	
Name:		D. O. B.:	
Sex:	Birth Type:		Breed:
Type:	○ Lamb ○ Ewe ○ Ram		Genotype:

Acquisition Details

Purchased From:		Date
Purchased address:		

Purpose of Acquisition: ○Breeding   ○Raising   ○Sale   ○Other:

Date Applied for Identifiers: | Purchase Price:

Weight Details	At Birth	Day 30	Day 60	6 Months	9 Month	1 Year

Pedigree Details

Sheep's Picture

Notes_____

SHEEP RECORD

Date Added:		Sheep Id:	
Name:		D. O. B.:	
Sex:	Birth Type:	Breed:	
Type:	○ Lamb ○ Ewe ○ Ram	Genotype:	

Acquisition Details

Purchased From:		Date
Purchased address:		

Purpose of Acquisition: ○Breeding   ○Raising   ○Sale   ○Other:

Date Applied for Identifiers: _____ | Purchase Price:

Weight Details	At Birth	Day 30	Day 60	6 Months	9 Month	1 Year

Pedigree Details

Sheep's Picture

Notes_____

_____

SHEEP RECORD

Date Added:		Sheep Id:
Name:		D. O. B.:
Sex:	Birth Type:	Breed:
Type:	◯ Lamb ◯ Ewe ◯ Ram	Genotype:

Acquisition Details

Purchased From:	Date
Purchased address:	

Purpose of Acquisition: ◯Breeding   ◯Raising   ◯Sale   ◯Other:

Date Applied for Identifiers:	Purchase Price:

Weight Details	At Birth	Day 30	Day 60	6 Months	9 Month	1 Year

Pedigree Details

Sheep's Picture

Notes _____

_____

57

SHEEP RECORD

Date Added:	Sheep Id:
Name:	D. O. B.:

Sex:	Birth Type:		Breed:
Type:	○ Lamb ○ Ewe ○ Ram		Genotype:

Acquisition Details

Purchased From:	Date
Purchased address:	

Purpose of Acquisition: ○Breeding   ○Raising   ○Sale   ○Other:

Date Applied for Identifiers:	Purchase Price:

Weight Details	At Birth	Day 30	Day 60	6 Months	9 Month	1 Year

Pedigree Details

Sheep's Picture

Notes_____

_____

SHEEP RECORD

Date Added:	Sheep Id:
Name:	D. O. B.:

Sex:			Birth Type:			Breed:

Type:	○ Lamb	○ Ewe	○ Ram	Genotype:

Acquisition Details

Purchased From:	Date

Purchased address:

Purpose of Acquisition: ○Breeding   ○Raising   ○Sale   ○Other:

Date Applied for Identifiers:	Purchase Price:

Weight Details	At Birth	Day 30	Day 60	6 Months	9 Month	1 Year

Pedigree Details

Sheep's Picture

Notes_____

_____

SHEEP RECORD

Date Added:	Sheep Id:
Name:	D. O. B.:

Sex:	Birth Type:	Breed:
Type: ○ Lamb ○ Ewe ○ Ram		Genotype:

Acquisition Details

Purchased From:	Date
Purchased address:	

Purpose of Acquisition: ○ Breeding   ○ Raising   ○ Sale   ○ Other:

Date Applied for Identifiers:	Purchase Price:

Weight Details	At Birth	Day 30	Day 60	6 Months	9 Month	1 Year

Pedigree Details

Sheep's Picture

Notes_____

_____

SHEEP RECORD

Date Added:		Sheep Id:
Name:		D. O. B.:
Sex:	Birth Type:	Breed:
Type:	○ Lamb ○ Ewe ○ Ram	Genotype:

Acquisition Details

Purchased From:	Date
Purchased address:	

Purpose of Acquisition: ○Breeding ○Raising ○Sale ○Other:

Date Applied for Identifiers: ____ Purchase Price: ____

Weight Details	At Birth	Day 30	Day 60	6 Months	9 Month	1 Year

Pedigree Details

Sheep's Picture

Notes_____

_____

SHEEP RECORD

Date Added:		Sheep Id:	
Name:		D. O. B.:	
Sex:	Birth Type:		Breed:
Type:	○ Lamb ○ Ewe ○ Ram		Genotype:

Acquisition Details

Purchased From:		Date
Purchased address:		

Purpose of Acquisition: ○ Breeding   ○ Raising   ○ Sale   ○ Other:

Date Applied for Identifiers:	Purchase Price:

Weight Details	At Birth	Day 30	Day 60	6 Months	9 Month	1 Year

Pedigree Details

Sheep's Picture

Notes_____

_____

SHEEP RECORD

Date Added:	Sheep Id:
Name:	D. O. B.:

Sex:	Birth Type:	Breed:

Type:	◯ Lamb ◯ Ewe ◯ Ram	Genotype:

Acquisition Details

Purchased From:	Date

Purchased address:

Purpose of Acquisition: ◯Breeding   ◯Raising   ◯Sale   ◯Other:

Date Applied for Identifiers: Purchase Price:

Weight Details	At Birth	Day 30	Day 60	6 Months	9 Month	1 Year

Pedigree Details

Sheep's Picture

Notes _____

_____

SHEEP RECORD

Date Added:	Sheep Id:
Name:	D. O. B.:

Sex:	Birth Type:		Breed:
Type:	○ Lamb ○ Ewe ○ Ram		Genotype:

Acquisition Details

Purchased From:	Date
Purchased address:	

Purpose of Acquisition: ○ Breeding   ○ Raising   ○ Sale   ○ Other:

Date Applied for Identifiers:	Purchase Price:

Weight Details	At Birth	Day 30	Day 60	6 Months	9 Month	1 Year

Pedigree Details

Sheep's Picture

Notes_____

_____

SHEEP RECORD

Date Added:		Sheep Id:	
Name:		D. O. B.:	
Sex:	Birth Type:		Breed:
Type:	◯ Lamb ◯ Ewe ◯ Ram		Genotype:

Acquisition Details

Purchased From:		Date
Purchased address:		

Purpose of Acquisition: ◯Breeding   ◯Raising   ◯Sale   ◯Other:

| Date Applied for Identifiers: | | Purchase Price: |

Weight Details	At Birth	Day 30	Day 60	6 Months	9 Month	1 Year

Pedigree Details

Sheep's Picture

Notes_____

SHEEP RECORD

Date Added:	Sheep Id:
Name:	D. O. B.:
Sex:	Birth Type:	Breed:
Type:	○ Lamb ○ Ewe ○ Ram	Genotype:

Acquisition Details

Purchased From:	Date
Purchased address:	

Purpose of Acquisition: ○Breeding   ○Raising   ○Sale   ○Other:

Date Applied for Identifiers: | Purchase Price:

Weight Details	At Birth	Day 30	Day 60	6 Months	9 Month	1 Year

Pedigree Details

Sheep's Picture

Notes_____

_____

SHEEP RECORD

Date Added:		Sheep Id:	
Name:		D. O. B.:	
Sex:	Birth Type:		Breed:
Type:	○ Lamb ○ Ewe ○ Ram		Genotype:

Acquisition Details

Purchased From:		Date
Purchased address:		

Purpose of Acquisition: ○Breeding   ○Raising   ○Sale   ○Other:

Date Applied for Identifiers: _____ Purchase Price: _____

Weight Details	At Birth	Day 30	Day 60	6 Months	9 Month	1 Year

Pedigree Details

Sheep's Picture

Notes_____

_____

SHEEP RECORD

Date Added:		Sheep Id:
Name:		D. O. B.:
Sex:	Birth Type:	Breed:
Type:	◯ Lamb ◯ Ewe ◯ Ram	Genotype:

Acquisition Details

| Purchased From: | Date |
| Purchased address: | |

Purpose of Acquisition: ◯Breeding   ◯Raising   ◯Sale   ◯Other:

Date Applied for Identifiers: | Purchase Price:

Weight Details	At Birth	Day 30	Day 60	6 Months	9 Month	1 Year

Pedigree Details

Sheep's Picture

Notes_____

_____

SHEEP RECORD

Date Added:		Sheep Id:	
Name:		D. O. B.:	
Sex:	Birth Type:		Breed:
Type:	○ Lamb ○ Ewe ○ Ram		Genotype:

Acquisition Details

Purchased From:		Date
Purchased address:		

Purpose of Acquisition: ○Breeding ○Raising ○Sale ○Other:

Date Applied for Identifiers: | Purchase Price:

Weight Details	At Birth	Day 30	Day 60	6 Months	9 Month	1 Year

Pedigree Details

Sheep's Picture

Notes_____

SHEEP RECORD

Date Added:	Sheep Id:
Name:	D. O. B.:

Sex:	Birth Type:	Breed:
Type: ○ Lamb ○ Ewe ○ Ram		Genotype:

Acquisition Details

Purchased From:	Date
Purchased address:	

Purpose of Acquisition: ○ Breeding   ○ Raising   ○ Sale   ○ Other:

Date Applied for Identifiers:	Purchase Price:

Weight Details	At Birth	Day 30	Day 60	6 Months	9 Month	1 Year

Pedigree Details

Sheep's Picture

Notes_____

_____

SHEEP RECORD

Date Added:	Sheep Id:
Name:	D. O. B.:

Sex:	Birth Type:	Breed:

Type:	○ Lamb ○ Ewe ○ Ram	Genotype:

Acquisition Details

Purchased From:	Date
Purchased address:	

Purpose of Acquisition: ○Breeding ○Raising ○Sale ○Other:

Date Applied for Identifiers:	Purchase Price:

Weight Details	At Birth	Day 30	Day 60	6 Months	9 Month	1 Year

Pedigree Details

Sheep's Picture

Notes_____

_____

BREEDING RECORD

Name:	Sheep Id.:
Mating Date:	Lambing Date:
Breed:	Number of Births:
Notes:	

Name:	Sheep Id.:
Mating Date:	Lambing Date:
Breed:	Number of Births:
Notes:	

Name:	Sheep Id.:
Mating Date:	Lambing Date:
Breed:	Number of Births:
Notes:	

Name:	Sheep Id.:
Mating Date:	Lambing Date:
Breed:	Number of Births:
Notes:	

BREEDING RECORD

Name:	Sheep Id.:
Mating Date:	Lambing Date:
Breed:	Number of Births:

Notes:

Name:	Sheep Id.:
Mating Date:	Lambing Date:
Breed:	Number of Births:

Notes:

Name:	Sheep Id.:
Mating Date:	Lambing Date:
Breed:	Number of Births:

Notes:

Name:	Sheep Id.:
Mating Date:	Lambing Date:
Breed:	Number of Births:

Notes:

BREEDING RECORD

Name:	Sheep Id.:
Mating Date:	Lambing Date:
Breed:	Number of Births:

Notes:

Name:	Sheep Id.:
Mating Date:	Lambing Date:
Breed:	Number of Births:

Notes:

Name:	Sheep Id.:
Mating Date:	Lambing Date:
Breed:	Number of Births:

Notes:

Name:	Sheep Id.:
Mating Date:	Lambing Date:
Breed:	Number of Births:

Notes:

BREEDING RECORD

Name:	Sheep Id.:
Mating Date:	Lambing Date:
Breed:	Number of Births:

Notes:

Name:	Sheep Id.:
Mating Date:	Lambing Date:
Breed:	Number of Births:

Notes:

Name:	Sheep Id.:
Mating Date:	Lambing Date:
Breed:	Number of Births:

Notes:

Name:	Sheep Id.:
Mating Date:	Lambing Date:
Breed:	Number of Births:

Notes:

BREEDING RECORD

Name:	Sheep Id.:
Mating Date:	Lambing Date:
Breed:	Number of Births:
Notes:	

Name:	Sheep Id.:
Mating Date:	Lambing Date:
Breed:	Number of Births:
Notes:	

Name:	Sheep Id.:
Mating Date:	Lambing Date:
Breed:	Number of Births:
Notes:	

Name:	Sheep Id.:
Mating Date:	Lambing Date:
Breed:	Number of Births:
Notes:	

BREEDING RECORD

Name:	Sheep Id.:
Mating Date:	Lambing Date:
Breed:	Number of Births:

Notes:

Name:	Sheep Id.:
Mating Date:	Lambing Date:
Breed:	Number of Births:

Notes:

Name:	Sheep Id.:
Mating Date:	Lambing Date:
Breed:	Number of Births:

Notes:

Name:	Sheep Id.:
Mating Date:	Lambing Date:
Breed:	Number of Births:

Notes:

BREEDING RECORD

Name:	Sheep Id.:
Mating Date:	Lambing Date:
Breed:	Number of Births:
Notes:	

Name:	Sheep Id.:
Mating Date:	Lambing Date:
Breed:	Number of Births:
Notes:	

Name:	Sheep Id.:
Mating Date:	Lambing Date:
Breed:	Number of Births:
Notes:	

Name:	Sheep Id.:
Mating Date:	Lambing Date:
Breed:	Number of Births:
Notes:	

BREEDING RECORD

Name:	Sheep Id.:
Mating Date:	Lambing Date:
Breed:	Number of Births:

Notes:

Name:	Sheep Id.:
Mating Date:	Lambing Date:
Breed:	Number of Births:

Notes:

Name:	Sheep Id.:
Mating Date:	Lambing Date:
Breed:	Number of Births:

Notes:

Name:	Sheep Id.:
Mating Date:	Lambing Date:
Breed:	Number of Births:

Notes:

BREEDING RECORD

Name:	Sheep Id.:
Mating Date:	Lambing Date:
Breed:	Number of Births:

Notes:

Name:	Sheep Id.:
Mating Date:	Lambing Date:
Breed:	Number of Births:

Notes:

Name:	Sheep Id.:
Mating Date:	Lambing Date:
Breed:	Number of Births:

Notes:

Name:	Sheep Id.:
Mating Date:	Lambing Date:
Breed:	Number of Births:

Notes:

BREEDING RECORD

Name:	Sheep Id.:
Mating Date:	Lambing Date:
Breed:	Number of Births:

Notes:

Name:	Sheep Id.:
Mating Date:	Lambing Date:
Breed:	Number of Births:

Notes:

Name:	Sheep Id.:
Mating Date:	Lambing Date:
Breed:	Number of Births:

Notes:

Name:	Sheep Id.:
Mating Date:	Lambing Date:
Breed:	Number of Births:

Notes:

BREEDING RECORD

Name:	Sheep Id.:
Mating Date:	Lambing Date:
Breed:	Number of Births:
Notes:	

Name:	Sheep Id.:
Mating Date:	Lambing Date:
Breed:	Number of Births:
Notes:	

Name:	Sheep Id.:
Mating Date:	Lambing Date:
Breed:	Number of Births:
Notes:	

Name:	Sheep Id.:
Mating Date:	Lambing Date:
Breed:	Number of Births:
Notes:	

BREEDING RECORD

Name:	Sheep Id.:
Mating Date:	Lambing Date:
Breed:	Number of Births:

Notes:

Name:	Sheep Id.:
Mating Date:	Lambing Date:
Breed:	Number of Births:

Notes:

Name:	Sheep Id.:
Mating Date:	Lambing Date:
Breed:	Number of Births:

Notes:

Name:	Sheep Id.:
Mating Date:	Lambing Date:
Breed:	Number of Births:

Notes:

BREEDING RECORD

Name:	Sheep Id.:
Mating Date:	Lambing Date:
Breed:	Number of Births:

Notes:

Name:	Sheep Id.:
Mating Date:	Lambing Date:
Breed:	Number of Births:

Notes:

Name:	Sheep Id.:
Mating Date:	Lambing Date:
Breed:	Number of Births:

Notes:

Name:	Sheep Id.:
Mating Date:	Lambing Date:
Breed:	Number of Births:

Notes:

BREEDING RECORD

Name:	Sheep Id.:
Mating Date:	Lambing Date:
Breed:	Number of Births:
Notes:	

Name:	Sheep Id.:
Mating Date:	Lambing Date:
Breed:	Number of Births:
Notes:	

Name:	Sheep Id.:
Mating Date:	Lambing Date:
Breed:	Number of Births:
Notes:	

Name:	Sheep Id.:
Mating Date:	Lambing Date:
Breed:	Number of Births:
Notes:	

BREEDING RECORD

Name:	Sheep Id.:
Mating Date:	Lambing Date:
Breed:	Number of Births:
Notes:	

Name:	Sheep Id.:
Mating Date:	Lambing Date:
Breed:	Number of Births:
Notes:	

Name:	Sheep Id.:
Mating Date:	Lambing Date:
Breed:	Number of Births:
Notes:	

Name:	Sheep Id.:
Mating Date:	Lambing Date:
Breed:	Number of Births:
Notes:	

BREEDING RECORD

Name:	Sheep Id.:
Mating Date:	Lambing Date:
Breed:	Number of Births:

Notes:

Name:	Sheep Id.:
Mating Date:	Lambing Date:
Breed:	Number of Births:

Notes:

Name:	Sheep Id.:
Mating Date:	Lambing Date:
Breed:	Number of Births:

Notes:

Name:	Sheep Id.:
Mating Date:	Lambing Date:
Breed:	Number of Births:

Notes:

BREEDING RECORD

Name:	Sheep Id.:
Mating Date:	Lambing Date:
Breed:	Number of Births:

Notes:

Name:	Sheep Id.:
Mating Date:	Lambing Date:
Breed:	Number of Births:

Notes:

Name:	Sheep Id.:
Mating Date:	Lambing Date:
Breed:	Number of Births:

Notes:

Name:	Sheep Id.:
Mating Date:	Lambing Date:
Breed:	Number of Births:

Notes:

BREEDING RECORD

Name:	Sheep Id.:
Mating Date:	Lambing Date:
Breed:	Number of Births:
Notes:	

Name:	Sheep Id.:
Mating Date:	Lambing Date:
Breed:	Number of Births:
Notes:	

Name:	Sheep Id.:
Mating Date:	Lambing Date:
Breed:	Number of Births:
Notes:	

Name:	Sheep Id.:
Mating Date:	Lambing Date:
Breed:	Number of Births:
Notes:	

BREEDING RECORD

Name:	Sheep Id.:
Mating Date:	Lambing Date:
Breed:	Number of Births:
Notes:	

Name:	Sheep Id.:
Mating Date:	Lambing Date:
Breed:	Number of Births:
Notes:	

Name:	Sheep Id.:
Mating Date:	Lambing Date:
Breed:	Number of Births:
Notes:	

Name:	Sheep Id.:
Mating Date:	Lambing Date:
Breed:	Number of Births:
Notes:	

BREEDING RECORD

Name:	Sheep Id.:
Mating Date:	Lambing Date:
Breed:	Number of Births:

Notes:

Name:	Sheep Id.:
Mating Date:	Lambing Date:
Breed:	Number of Births:

Notes:

Name:	Sheep Id.:
Mating Date:	Lambing Date:
Breed:	Number of Births:

Notes:

Name:	Sheep Id.:
Mating Date:	Lambing Date:
Breed:	Number of Births:

Notes:

LAMBING RECORD

Date	Sheep Id.	Breed	Birth Type	Weight

Notes

_____

_____

_____

_____

_____

LAMBING RECORD

Date	Sheep Id.	Breed	Birth Type	Weight

Notes

_____

_____

_____

_____

_____

LAMBING RECORD

Date	Sheep Id.	Breed	Birth Type	Weight

Notes

_____

_____

_____

_____

_____

LAMBING RECORD

Date	Sheep Id.	Breed	Birth Type	Weight

Notes

_____

_____

_____

_____

_____

LAMBING RECORD

Date	Sheep Id.	Breed	Birth Type	Weight

Notes

_____

_____

_____

_____

_____

LAMBING RECORD

Date	Sheep Id.	Breed	Birth Type	Weight

Notes

LAMBING RECORD

Date	Sheep Id.	Breed	Birth Type	Weight

Notes

_____

_____

_____

_____

_____

LAMBING RECORD

Date	Sheep Id.	Breed	Birth Type	Weight

Notes

_____

_____

_____

_____

_____

LAMBING RECORD

Date	Sheep Id.	Breed	Birth Type	Weight

Notes

_____

_____

_____

_____

_____

LAMBING RECORD

Date	Sheep Id.	Breed	Birth Type	Weight

Notes

_____

_____

_____

_____

_____

HEALTH RECORD

Name:				Sheep Id:
Date:	Symptoms			Actions
Time:				
Age				

Treatment Given				Response
Product				
Dose				
Duration				
Comments				

Name:				Sheep Id:
Date:	Symptoms			Actions
Time:				
Age				

Treatment Given				Response
Product				
Dose				
Duration				
Comments				

Name:				Sheep Id:
Date:	Symptoms			Actions
Time:				
Age				

Treatment Given				Response
Product				
Dose				
Duration				
Comments				

HEALTH RECORD

Name:				Sheep Id:
Date:	Symptoms			Actions
Time:				
Age				

Treatment Given				Response
Product				
Dose				
Duration				
Comments				

Name:				Sheep Id:
Date:	Symptoms			Actions
Time:				
Age				

Treatment Given				Response
Product				
Dose				
Duration				
Comments				

Name:				Sheep Id:
Date:	Symptoms			Actions
Time:				
Age				

Treatment Given				Response
Product				
Dose				
Duration				
Comments				

HEALTH RECORD

Name:		Sheep Id:
Date:	Symptoms	Actions
Time:		
Age		

Treatment Given				Response
Product				
Dose				
Duration				
Comments				

Name:		Sheep Id:
Date:	Symptoms	Actions
Time:		
Age		

Treatment Given				Response
Product				
Dose				
Duration				
Comments				

Name:		Sheep Id:
Date:	Symptoms	Actions
Time:		
Age		

Treatment Given				Response
Product				
Dose				
Duration				
Comments				

HEALTH RECORD

Name:			Sheep Id:	
Date:	Symptoms		Actions	
Time:				
Age				

Treatment Given				Response
Product				
Dose				
Duration				
Comments				

Name:			Sheep Id:	
Date:	Symptoms		Actions	
Time:				
Age				

Treatment Given				Response
Product				
Dose				
Duration				
Comments				

Name:			Sheep Id:	
Date:	Symptoms		Actions	
Time:				
Age				

Treatment Given				Response
Product				
Dose				
Duration				
Comments				

HEALTH RECORD

Name:				Sheep Id:
Date:	Symptoms			Actions
Time:				
Age				
Treatment Given				Response
Product				
Dose				
Duration				
Comments				

Name:				Sheep Id:
Date:	Symptoms			Actions
Time:				
Age				
Treatment Given				Response
Product				
Dose				
Duration				
Comments				

Name:				Sheep Id:
Date:	Symptoms			Actions
Time:				
Age				
Treatment Given				Response
Product				
Dose				
Duration				
Comments				

HEALTH RECORD

Name:				Sheep Id:
Date:	Symptoms			Actions
Time:				
Age				
Treatment Given				Response
Product				
Dose				
Duration				
Comments				

Name:				Sheep Id:
Date:	Symptoms			Actions
Time:				
Age				
Treatment Given				Response
Product				
Dose				
Duration				
Comments				

Name:				Sheep Id:
Date:	Symptoms			Actions
Time:				
Age				
Treatment Given				Response
Product				
Dose				
Duration				
Comments				

HEALTH RECORD

Name:				Sheep Id:
Date:	Symptoms			Actions
Time:				
Age				

Treatment Given				Response
Product				
Dose				
Duration				
Comments				

Name:				Sheep Id:
Date:	Symptoms			Actions
Time:				
Age				

Treatment Given				Response
Product				
Dose				
Duration				
Comments				

Name:				Sheep Id:
Date:	Symptoms			Actions
Time:				
Age				

Treatment Given				Response
Product				
Dose				
Duration				
Comments				

HEALTH RECORD

Name:			Sheep Id:	
Date:	Symptoms		Actions	
Time:				
Age				

Treatment Given				Response
Product				
Dose				
Duration				
Comments				

Name:			Sheep Id:	
Date:	Symptoms		Actions	
Time:				
Age				

Treatment Given				Response
Product				
Dose				
Duration				
Comments				

Name:			Sheep Id:	
Date:	Symptoms		Actions	
Time:				
Age				

Treatment Given				Response
Product				
Dose				
Duration				
Comments				

HEALTH RECORD

Name:		Sheep Id:
Date:	Symptoms	Actions
Time:		
Age		

Treatment Given				Response
Product				
Dose				
Duration				
Comments				

Name:		Sheep Id:
Date:	Symptoms	Actions
Time:		
Age		

Treatment Given				Response
Product				
Dose				
Duration				
Comments				

Name:		Sheep Id:
Date:	Symptoms	Actions
Time:		
Age		

Treatment Given				Response
Product				
Dose				
Duration				
Comments				

HEALTH RECORD

Name:				Sheep Id:
Date:	Symptoms			Actions
Time:				
Age				

Treatment Given				Response
Product				
Dose				
Duration				
Comments				

Name:				Sheep Id:
Date:	Symptoms			Actions
Time:				
Age				

Treatment Given				Response
Product				
Dose				
Duration				
Comments				

Name:				Sheep Id:
Date:	Symptoms			Actions
Time:				
Age				

Treatment Given				Response
Product				
Dose				
Duration				
Comments				

HEALTH RECORD

Name:				Sheep Id:
Date:	Symptoms			Actions
Time:				
Age				
	Treatment Given			Response
Product				
Dose				
Duration				
Comments				

Name:				Sheep Id:
Date:	Symptoms			Actions
Time:				
Age				
	Treatment Given			Response
Product				
Dose				
Duration				
Comments				

Name:				Sheep Id:
Date:	Symptoms			Actions
Time:				
Age				
	Treatment Given			Response
Product				
Dose				
Duration				
Comments				

HEALTH RECORD

Name:				Sheep Id:
Date:	Symptoms			Actions
Time:				
Age				

Treatment Given				Response
Product				
Dose				
Duration				
Comments				

Name:				Sheep Id:
Date:	Symptoms			Actions
Time:				
Age				

Treatment Given				Response
Product				
Dose				
Duration				
Comments				

Name:				Sheep Id:
Date:	Symptoms			Actions
Time:				
Age				

Treatment Given				Response
Product				
Dose				
Duration				
Comments				

HEALTH RECORD

Name:				Sheep Id:
Date:	Symptoms			Actions
Time:				
Age				

Treatment Given				Response
Product				
Dose				
Duration				
Comments				

Name:				Sheep Id:
Date:	Symptoms			Actions
Time:				
Age				

Treatment Given				Response
Product				
Dose				
Duration				
Comments				

Name:				Sheep Id:
Date:	Symptoms			Actions
Time:				
Age				

Treatment Given				Response
Product				
Dose				
Duration				
Comments				

HEALTH RECORD

Name:				Sheep Id:
Date:	Symptoms			Actions
Time:				
Age				

Treatment Given				Response
Product				
Dose				
Duration				
Comments				

Name:				Sheep Id:
Date:	Symptoms			Actions
Time:				
Age				

Treatment Given				Response
Product				
Dose				
Duration				
Comments				

Name:				Sheep Id:
Date:	Symptoms			Actions
Time:				
Age				

Treatment Given				Response
Product				
Dose				
Duration				
Comments				

HEALTH RECORD

Name:				Sheep Id:
Date:	Symptoms			Actions
Time:				
Age				

Treatment Given				Response
Product				
Dose				
Duration				
Comments				

Name:				Sheep Id:
Date:	Symptoms			Actions
Time:				
Age				

Treatment Given				Response
Product				
Dose				
Duration				
Comments				

Name:				Sheep Id:
Date:	Symptoms			Actions
Time:				
Age				

Treatment Given				Response
Product				
Dose				
Duration				
Comments				

HEALTH RECORD

Name:		Sheep Id:
Date:	Symptoms	Actions
Time:		
Age		

Treatment Given			Response
Product			
Dose			
Duration			
Comments			

Name:		Sheep Id:
Date:	Symptoms	Actions
Time:		
Age		

Treatment Given			Response
Product			
Dose			
Duration			
Comments			

Name:		Sheep Id:
Date:	Symptoms	Actions
Time:		
Age		

Treatment Given			Response
Product			
Dose			
Duration			
Comments			

HEALTH RECORD

Name:				Sheep Id:
Date:		Symptoms		Actions
Time:				
Age				

	Treatment Given			Response
Product				
Dose				
Duration				
Comments				

Name:				Sheep Id:
Date:		Symptoms		Actions
Time:				
Age				

	Treatment Given			Response
Product				
Dose				
Duration				
Comments				

Name:				Sheep Id:
Date:		Symptoms		Actions
Time:				
Age				

	Treatment Given			Response
Product				
Dose				
Duration				
Comments				

HEALTH RECORD

Name:		Sheep Id:
Date:	Symptoms	Actions
Time:		
Age		

	Treatment Given			Response
Product				
Dose				
Duration				
Comments				

Name:		Sheep Id:
Date:	Symptoms	Actions
Time:		
Age		

	Treatment Given			Response
Product				
Dose				
Duration				
Comments				

Name:		Sheep Id:
Date:	Symptoms	Actions
Time:		
Age		

	Treatment Given			Response
Product				
Dose				
Duration				
Comments				

HEALTH RECORD

Name:				Sheep Id:
Date:	Symptoms			Actions
Time:				
Age				

Treatment Given				Response
Product				
Dose				
Duration				
Comments				

Name:				Sheep Id:
Date:	Symptoms			Actions
Time:				
Age				

Treatment Given				Response
Product				
Dose				
Duration				
Comments				

Name:				Sheep Id:
Date:	Symptoms			Actions
Time:				
Age				

Treatment Given				Response
Product				
Dose				
Duration				
Comments				

HEALTH RECORD

Name:				Sheep Id:
Date:	Symptoms			Actions
Time:				
Age				

Treatment Given				Response
Product				
Dose				
Duration				
Comments				

Name:				Sheep Id:
Date:	Symptoms			Actions
Time:				
Age				

Treatment Given				Response
Product				
Dose				
Duration				
Comments				

Name:				Sheep Id:
Date:	Symptoms			Actions
Time:				
Age				

Treatment Given				Response
Product				
Dose				
Duration				
Comments				

DEATH RECORD

Date:	Name:	
Age:	Sheep Id.:	Cause of Death:
Summary:		

Date:	Name:	
Age:	Sheep Id.:	Cause of Death:
Summary:		

Date:	Name:	
Age:	Sheep Id.:	Cause of Death:
Summary:		

Date:	Name:	
Age:	Sheep Id.:	Cause of Death:
Summary:		

Date:	Name:	
Age:	Sheep Id.:	Cause of Death:
Summary:		

DEATH RECORD

Date:	Name:	
Age:	Sheep Id.:	Cause of Death:

Summary:

Date:	Name:	
Age:	Sheep Id.:	Cause of Death:

Summary:

Date:	Name:	
Age:	Sheep Id.:	Cause of Death:

Summary:

Date:	Name:	
Age:	Sheep Id.:	Cause of Death:

Summary:

Date:	Name:	
Age:	Sheep Id.:	Cause of Death:

Summary:

DEATH RECORD

Date: | **Name:**

Age: | **Sheep Id.:** | **Cause of Death:**

Summary:

Date: | **Name:**

Age: | **Sheep Id.:** | **Cause of Death:**

Summary:

Date: | **Name:**

Age: | **Sheep Id.:** | **Cause of Death:**

Summary:

Date: | **Name:**

Age: | **Sheep Id.:** | **Cause of Death:**

Summary:

Date: | **Name:**

Age: | **Sheep Id.:** | **Cause of Death:**

Summary:

DEATH RECORD

Date:	**Name:**	
Age:	**Sheep Id.:**	**Cause of Death:**

Summary:

Date:	**Name:**	
Age:	**Sheep Id.:**	**Cause of Death:**

Summary:

Date:	**Name:**	
Age:	**Sheep Id.:**	**Cause of Death:**

Summary:

Date:	**Name:**	
Age:	**Sheep Id.:**	**Cause of Death:**

Summary:

Date:	**Name:**	
Age:	**Sheep Id.:**	**Cause of Death:**

Summary:

DEATH RECORD

Date: **Name:**

Age: **Sheep Id.:**   **Cause of Death:**

Summary:

Date: **Name:**

Age: **Sheep Id.:**   **Cause of Death:**

Summary:

Date: **Name:**

Age: **Sheep Id.:**   **Cause of Death:**

Summary:

Date: **Name:**

Age: **Sheep Id.:**   **Cause of Death:**

Summary:

Date: **Name:**

Age: **Sheep Id.:**   **Cause of Death:**

Summary:

DEATH RECORD

Date:	Name:	
Age:	Sheep Id.:	Cause of Death:

Summary:

Date:	Name:	
Age:	Sheep Id.:	Cause of Death:

Summary:

Date:	Name:	
Age:	Sheep Id.:	Cause of Death:

Summary:

Date:	Name:	
Age:	Sheep Id.:	Cause of Death:

Summary:

Date:	Name:	
Age:	Sheep Id.:	Cause of Death:

Summary:

DEATH RECORD

Date: **Name:**

Age: **Sheep Id.:**   **Cause of Death:**

Summary:

Date: **Name:**

Age: **Sheep Id.:**   **Cause of Death:**

Summary:

Date: **Name:**

Age: **Sheep Id.:**   **Cause of Death:**

Summary:

Date: **Name:**

Age: **Sheep Id.:**   **Cause of Death:**

Summary:

Date: **Name:**

Age: **Sheep Id.:**   **Cause of Death:**

Summary:

DEATH RECORD

Date:	Name:	
Age:	Sheep Id.:	Cause of Death:

Summary:

Date:	Name:	
Age:	Sheep Id.:	Cause of Death:

Summary:

Date:	Name:	
Age:	Sheep Id.:	Cause of Death:

Summary:

Date:	Name:	
Age:	Sheep Id.:	Cause of Death:

Summary:

Date:	Name:	
Age:	Sheep Id.:	Cause of Death:

Summary:

DEATH RECORD

Date:	Name:	
Age:	Sheep Id.:	Cause of Death:

Summary:

Date:	Name:	
Age:	Sheep Id.:	Cause of Death:

Summary:

Date:	Name:	
Age:	Sheep Id.:	Cause of Death:

Summary:

Date:	Name:	
Age:	Sheep Id.:	Cause of Death:

Summary:

Date:	Name:	
Age:	Sheep Id.:	Cause of Death:

Summary:

DEATH RECORD

Date: **Name:**

Age: **Sheep Id.:**   **Cause of Death:**

Summary:

Date: **Name:**

Age: **Sheep Id.:**   **Cause of Death:**

Summary:

Date: **Name:**

Age: **Sheep Id.:**   **Cause of Death:**

Summary:

Date: **Name:**

Age: **Sheep Id.:**   **Cause of Death:**

Summary:

Date: **Name:**

Age: **Sheep Id.:**   **Cause of Death:**

Summary:

DEATH RECORD

Date:	Name:	
Age:	Sheep Id.:	Cause of Death:

Summary:

Date:	Name:	
Age:	Sheep Id.:	Cause of Death:

Summary:

Date:	Name:	
Age:	Sheep Id.:	Cause of Death:

Summary:

Date:	Name:	
Age:	Sheep Id.:	Cause of Death:

Summary:

Date:	Name:	
Age:	Sheep Id.:	Cause of Death:

Summary:

DEATH RECORD

Date: Name:

Age: Sheep Id.:   Cause of Death:

Summary:

Date: Name:

Age: Sheep Id.:   Cause of Death:

Summary:

Date: Name:

Age: Sheep Id.:   Cause of Death:

Summary:

Date: Name:

Age: Sheep Id.:   Cause of Death:

Summary:

Date: Name:

Age: Sheep Id.:   Cause of Death:

Summary:

DEATH RECORD

Date:	Name:	
Age:	Sheep Id.:	Cause of Death:

Summary:

Date:	Name:	
Age:	Sheep Id.:	Cause of Death:

Summary:

Date:	Name:	
Age:	Sheep Id.:	Cause of Death:

Summary:

Date:	Name:	
Age:	Sheep Id.:	Cause of Death:

Summary:

Date:	Name:	
Age:	Sheep Id.:	Cause of Death:

Summary:

DEATH RECORD

Date: Name:

Age: Sheep Id.:   Cause of Death:

Summary:

Date: Name:

Age: Sheep Id.:   Cause of Death:

Summary:

Date: Name:

Age: Sheep Id.:   Cause of Death:

Summary:

Date: Name:

Age: Sheep Id.:   Cause of Death:

Summary:

Date: Name:

Age: Sheep Id.:   Cause of Death:

Summary:

DEATH RECORD

Date:	Name:	
Age:	Sheep Id.:	Cause of Death:

Summary:

Date:	Name:	
Age:	Sheep Id.:	Cause of Death:

Summary:

Date:	Name:	
Age:	Sheep Id.:	Cause of Death:

Summary:

Date:	Name:	
Age:	Sheep Id.:	Cause of Death:

Summary:

Date:	Name:	
Age:	Sheep Id.:	Cause of Death:

Summary:

DEATH RECORD

Date: | **Name:**

Age: | **Sheep Id.:** | **Cause of Death:**

Summary:

Date: | **Name:**

Age: | **Sheep Id.:** | **Cause of Death:**

Summary:

Date: | **Name:**

Age: | **Sheep Id.:** | **Cause of Death:**

Summary:

Date: | **Name:**

Age: | **Sheep Id.:** | **Cause of Death:**

Summary:

Date: | **Name:**

Age: | **Sheep Id.:** | **Cause of Death:**

Summary:

DEATH RECORD

Date:	Name:	
Age:	Sheep Id.:	Cause of Death:

Summary:

Date:	Name:	
Age:	Sheep Id.:	Cause of Death:

Summary:

Date:	Name:	
Age:	Sheep Id.:	Cause of Death:

Summary:

Date:	Name:	
Age:	Sheep Id.:	Cause of Death:

Summary:

Date:	Name:	
Age:	Sheep Id.:	Cause of Death:

Summary:

DEATH RECORD

Date: | Name:

Age: | Sheep Id.: | Cause of Death:

Summary:

Date: | Name:

Age: | Sheep Id.: | Cause of Death:

Summary:

Date: | Name:

Age: | Sheep Id.: | Cause of Death:

Summary:

Date: | Name:

Age: | Sheep Id.: | Cause of Death:

Summary:

Date: | Name:

Age: | Sheep Id.: | Cause of Death:

Summary:

DEATH RECORD

Date:	Name:	
Age:	Sheep Id.:	Cause of Death:

Summary:

Date:	Name:	
Age:	Sheep Id.:	Cause of Death:

Summary:

Date:	Name:	
Age:	Sheep Id.:	Cause of Death:

Summary:

Date:	Name:	
Age:	Sheep Id.:	Cause of Death:

Summary:

Date:	Name:	
Age:	Sheep Id.:	Cause of Death:

Summary:

DEATH RECORD

Date:	Name:	
Age:	Sheep Id.:	Cause of Death:

Summary:

Date:	Name:	
Age:	Sheep Id.:	Cause of Death:

Summary:

Date:	Name:	
Age:	Sheep Id.:	Cause of Death:

Summary:

Date:	Name:	
Age:	Sheep Id.:	Cause of Death:

Summary:

Date:	Name:	
Age:	Sheep Id.:	Cause of Death:

Summary:

NOTES

NOTES

NOTES

NOTES

NOTES

NOTES

NOTES